# FUN FACTS ABOUT THE SOLAR SYSTEM

## Author: AIDEN

The Beautiful Mind Press
www.thebeautifulmindpress.com

FUN FACTS ABOUT THE SOLAR SYSTEM

Copyright © 2023

Written and Hand Illustrated by AIDEN

All rights reserved. No part of this book may be reproduced or transmitted in any form or by any means without written permission from the author.

Printed in the USA

The Beautiful Mind Press
www.TheBeautifulMindPress.com

> Our solar system is made up of a star, eight planets, and countless smaller bodies such as dwarf planets, asteroids, and comets

> Our solar system is the only one known to support life. So far, we only know of life on Earth

> Our solar system orbits the center of the Milky Way galaxy at about 515,000 mph (828,000 kph)

> It takes our solar system about 230 million years to complete one orbit around the galactic center

From https://solarsystem.nasa.gov/solar-system/our-solar-system/overview/

# SUN

Average surface temperature of the Sun is 9941 F

The Sun's light reaches the Earth in 8 minutes

Sun is 91.467 million miles from earth

# MERCURY

> It takes 88 EARTH days to circle around the Sun

> It spins so slowly, that its day is equal to 59 Earth days!

> It travels at the speed of 29 miles per second. It is the fastest planet

# EARTH

It is at a perfect distance from Sun, so not so HOT, and it has Water and Atmosphere to support LIFE

NOT PERFECTLY ROUND! Earth's diameter from North to South Pole is 12,714 kilometers (7,900 miles), while through the equator it is 12,756 kilometers (7,926 miles)

Earth takes 365.25 Days to go around the Sun

# MARS

It is RED because of IRON

Maybe MARS may support LIFE One day!!

If you weigh 100 lbs on earth, on Mars you will be only 38 lbs

# JUPITER

> It has 92 Moons (more maybe found)

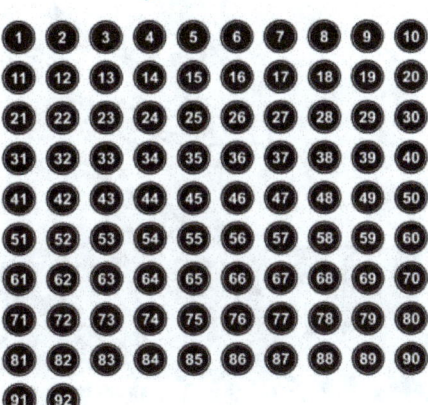

> It is the BIGGEST planet of our solar system

# SATURN

It is made of GAS! You cannot stand on it

It is made of Helium, that is in Birthday balloons

# URANUS

It is COLD. Average temperature is -353 F

Funny planet Uranus, spins SIDEWAYS!!!

# NEPTUNE

It is also called ICE GIANT

It takes 165 years to go around the Sun

# PLUTO

> It is a DWARF Planet. It is smaller than Earth's moon!

A dwarf planet is a small planetary-mass object that is in direct orbit of the Sun, smaller than any of the eight classical planets

# HAUMEA

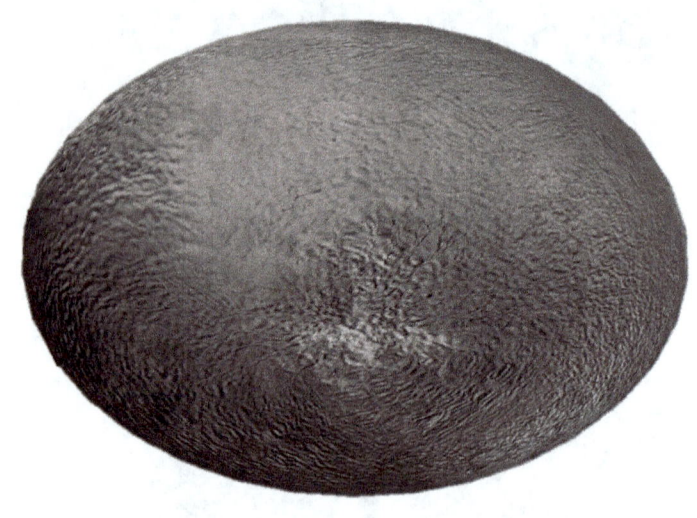

Haumea is a dwarf planet located beyond Neptune's orbit and has oval shape

# CERES

It is a dwarf planet in the asteroid belt between the orbits of Mars and Jupiter

# Makemake

It is a dwarf planet and the second-largest of what are known as the classical population of Kuiper belt objects, with a diameter approximately 60% that of Pluto

# Black Hole

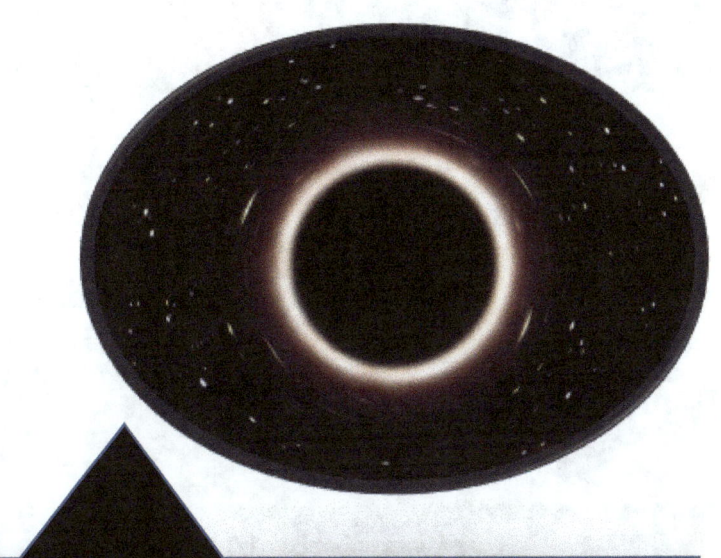

Nothing can escape Black Hole not even light because of the intense gravity; Gravity's speed is faster than that of Light!

DO NOT WORRY....Earth will not fall into a black hole because no black hole is close enough to the solar system for Earth to do that [According to NASA]

# Solar System Game

Travel the Solar System by using Dice and Pegs. If you step into The Black Hole, you have to start again!

# Find the Word Game

| | | | | | | | | | |
|---|---|---|---|---|---|---|---|---|---|
| I | R | V | E | N | U | S | M | Y | I | Y |
| E | A | R | T | H | K | P | U | I | O | C |
| M | B | L | O | H | O | T | N | R | T | A |
| A | L | I | G | H | T | O | I | U | Y | R |
| K | A | T | M | U | E | P | A | S | S | D |
| E | C | J | A | G | A | O | T | F | D | E |
| M | K | L | R | U | R | A | N | U | S | R |
| A | H | D | S | V | T | Y | O | O | G | O |
| K | O | Y | I | E | G | L | Z | P | H | C |
| E | L | O | O | N | L | Q | P | I | J | K |
| L | E | O | U | C | H | Y | L | K | P | B |
| O | L | I | O | S | A | T | U | R | N | A |
| J | U | P | I | T | E | R | T | M | C | C |
| N | E | P | T | U | N | E | O | E | G | U |
| A | V | T | W | D | Y | G | R | N | N | B |
| H | A | U | M | E | A | E | F | G | H | S |
| T | Z | H | F | E | S | T | I | F | L | W |
| A | I | D | E | N | E | L | L | I | E | M |

# Solar System Maze

See which places your spacecraft can visit in the Maze

www.ingramcontent.com/pod-product-compliance
Lightning Source LLC
Chambersburg PA
CBHW050323010526
44119CB00003B/83